AF888022

Transition Metal Complexes – Basics

Muhammad Jawwad Saif

All rights reserved.

ISBN: 978-1-64953-227-5

Absolute Author
Publishing House

1231 Williams Blvd
New Orleans, LA 70123

DEDICATION

I dedicate this book to my institute – Government College University Faisalabad, Pakistan

CONTENTS

1 INTRODUCTION

Transition metals involve the filling of d-orbitals or f-orbitals in their corresponding shells. In elements where the d-orbital is in the process of completion are called d-block elements. The elements where the f-orbital is in the process of completion are called f-block elements are inner transition elements.

Here are few important considerations:

(i) d-block elements lie in the center of the periodic table. These lie in between s-block and p-block elements.

(ii) d-block elements are present in the 4th, 5th, 6th and 7th periods.

4th period	(Sc to Zn)
5th period	(Y to Cd)
6th period	(La, Hf to Hg)
7th period	(Sc, Rf to Cn, incomplete)

(iii) d-block elements lie in IVB to VIIIB, IB, and IIB groups.

The elements which have incomplete d-orbital or f-orbital are called transition elements.

GENERAL PROPERTIES

The majority of the transition metals are similar in general properties. Below we are giving a brief overview.

We know that the atomic radii of elements generally decrease from left to right because of an increase in nuclear charge as we move in the period. The same is valid for transition metals.

Transition metals show variable oxidation states. Therefore, the ionic radii vary with the oxidation number. Generally, the ionic radii tend to decrease with an increase in formal charge on the metal atom. With similar formal charges, the ionic radii for various atoms depend on the corresponding metal atom's nuclear charge. The greater the nuclear charge, there is the decrease in the ionic radii.

All transition elements show strong metallic character. Transition metals show very high melting and boiling points. This characteristic is typical for metals. Aside from few exceptions, the density of transition metals usually increases along a period. This is due to decreasing atomic sizes and increased mass numbers.

IMPORTANT TERMS USED IN COORDINATION CHEMISTRY

There are terms very often used in describing the chemistry of coordination compounds.

Central Metal Atom

Ligands are attached to the central metal atom in a coordination complex. Mostly there is a single metal atom in a coordination complex molecule, but there are a number of metal complexes containing two or more than two metal atoms in a single metal complex molecule.

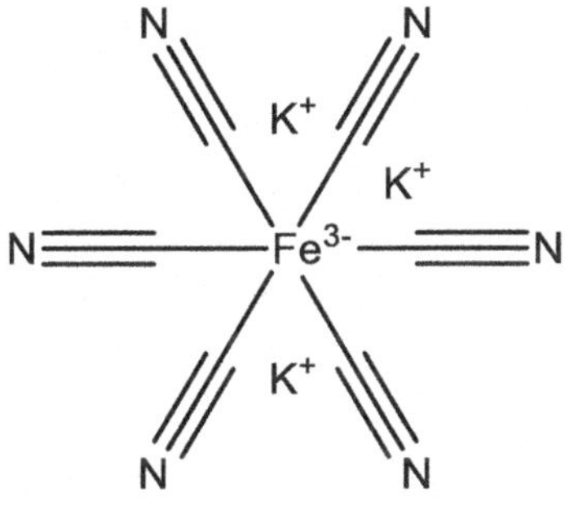

$K_3[Fe(CN)_6]$

potassium hexacyanoferrate(III)

In this example, iron is the central metal atom.

Ligands

These are the molecules attached to the central metal atom. Ligands are attached to the metal atom through a coordinate covalent bond. It means ligands are electron-donating species towards the central metal atom. Ligands can be negatively charged or neutral species, but these are electron-rich species.

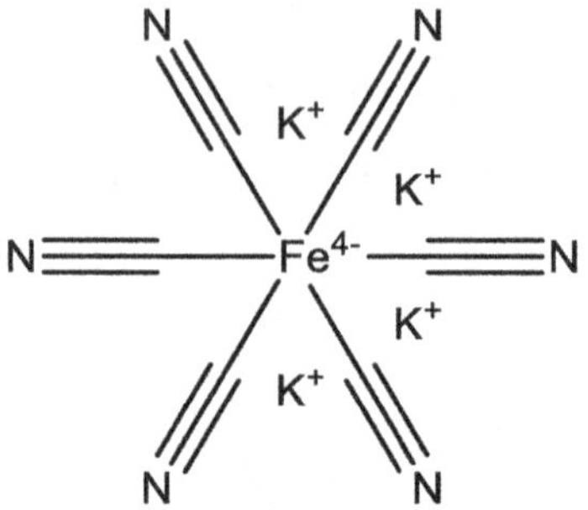

$K_4[Fe(CN)_6]$

potassium hexacyanoferrate(II)

In this example, six cyanide groups are coordinating with the central metal atoms and acting as ligands.

Coordination Number

Coordination number is the number of ligands attached with the central metal atom. If there are four ligands attached with the central metal atom, the coordination number is four. If there are six ligands attached with the central metal atom, the metal atom's coordination number is six.

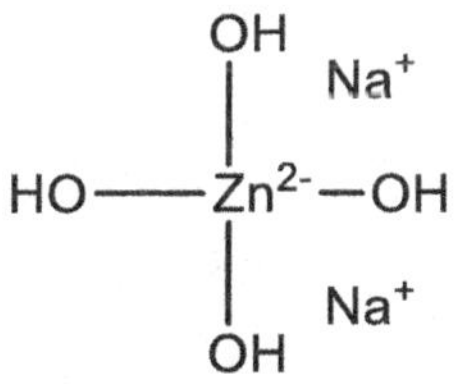

$Na_2[Zn(OH)_4]$

sodium tetrahydroxyzincate(II)

In this example, there are four hydroxy ligands attached to the zinc atom. Hence the coordination number of Zn is four in this complex.

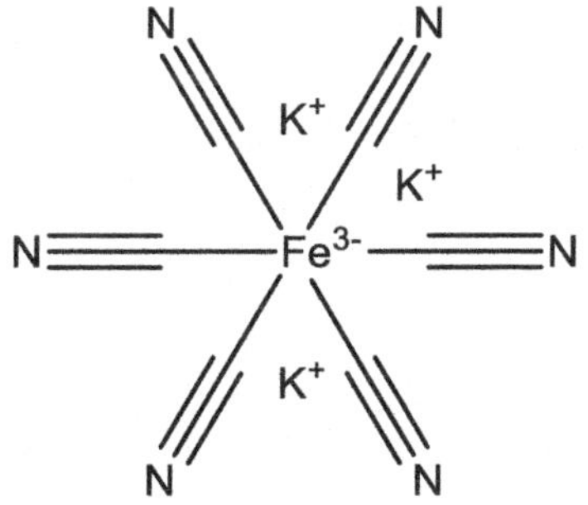

$K_3[Fe(CN)_6]$

potassium hexacyanoferrate(III)

In the above-given example, there are six CN groups attached to the iron atom. Hence, the coordination number of Fe (iron) is six in this complex.

Coordination Sphere

The coordination sphere is represented by a square bracket [ ] in writing the formula. It contains the central metal atom and the ligands.

For example, writing the formula of potassium hexacyanoferrate(III), we write it as

$K_3[Fe(CN)_6]$

The coordination sphere includes Fe as the central metal atom and CN ligands. Potassium is not written within the coordination sphere because it does not coordinate covalent bonds with the central metal atom. Potassium is considered an ionization sphere.

Ligands can be classified as

1. Monodentate Ligands
2. Polydentate Ligands

Monodentate Ligands

Monodentate ligands are those ligand species that have only a single coordination site available.

For example,

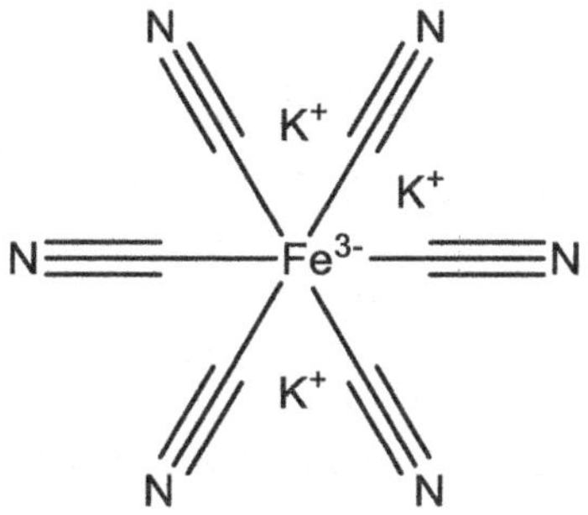

$K_3[Fe(CN)_6]$

potassium hexacyanoferrate(III)

In this complex, the CN ligands have only a single site to coordinate with the central metal atom.

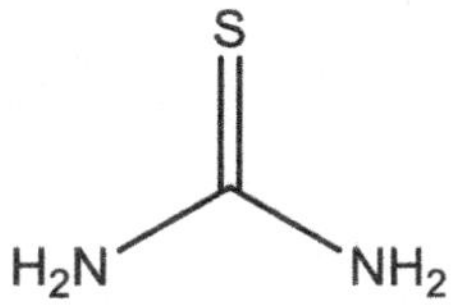

thiourea

NH_3

ammonia

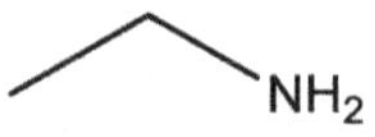

ethylamine

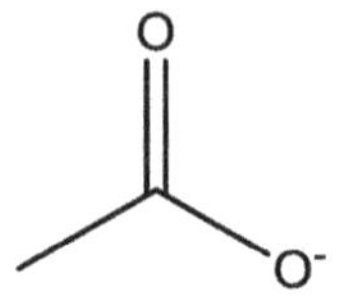

acetate

Polydentate Ligands

Ligands are carrying more than one coordinating site available are called polydentate ligands.

It can be further classified as

(i) Bidentate ligands

(ii) Tridentate ligands

(iii) Tetradentate ligands

(iv) Pentadentate ligands

(v) Hexadentate ligands

(i) Bidentate ligands

These ligands have two coordinating sites available on their structure. Examples include:

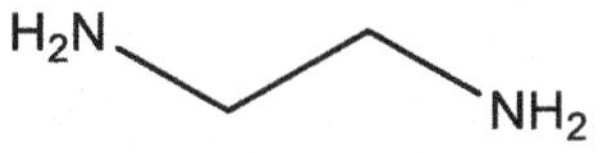

ethane-1,2-diamine

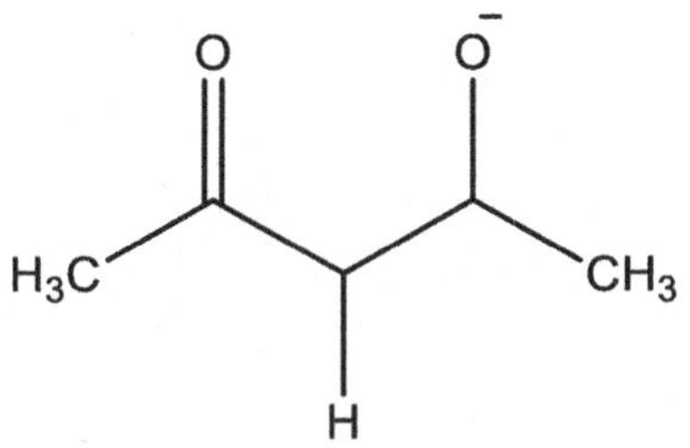

acetylacetonate

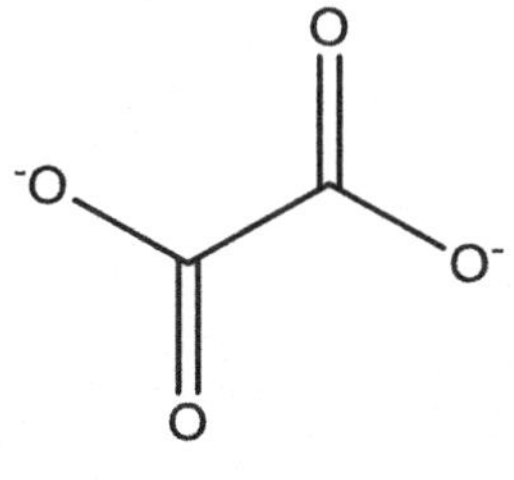

oxalate ion

Tridentate Ligands

These ligands are having three coordination sites available in their structure to coordinate with the central metal atom.

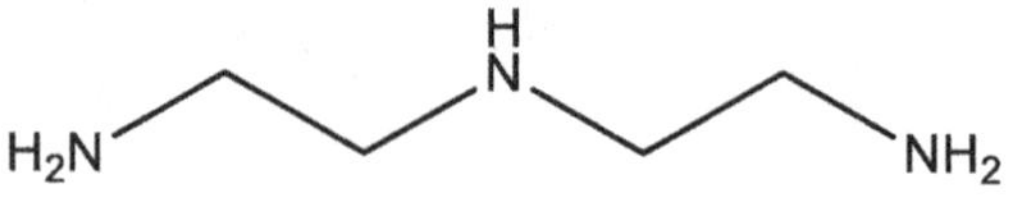

N^1-(2-aminoethyl)ethane-1,2-diamine

Tetradentate Ligands

These ligands are having four coordination sites available on a single ligand species.

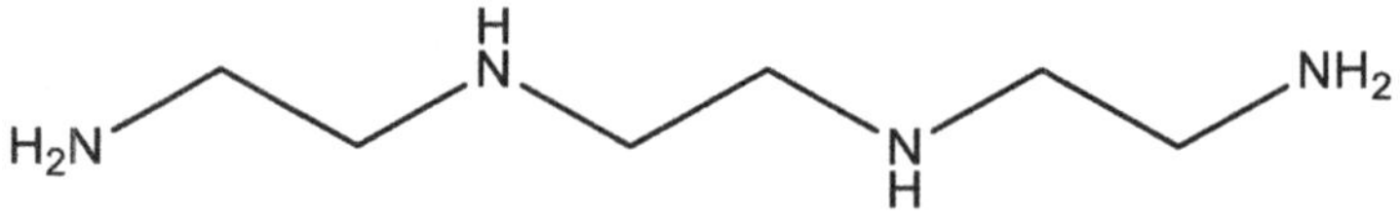

$N^1,N^{1'}$-(ethane-1,2-diyl)bis(ethane-1,2-diamine)

Pentadentate Ligands

If a single ligand carries five donor atoms in its structure, such a ligand is a pentadentate ligand.

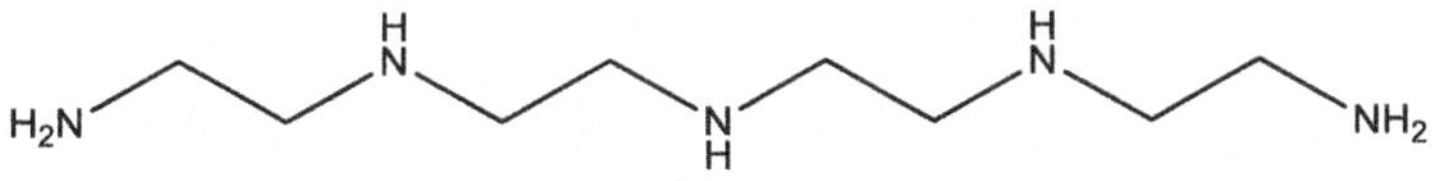

N^1-(2-aminoethyl)-N^2-(2-((2-aminoethyl)amino)ethyl)ethane-1,2-diamine

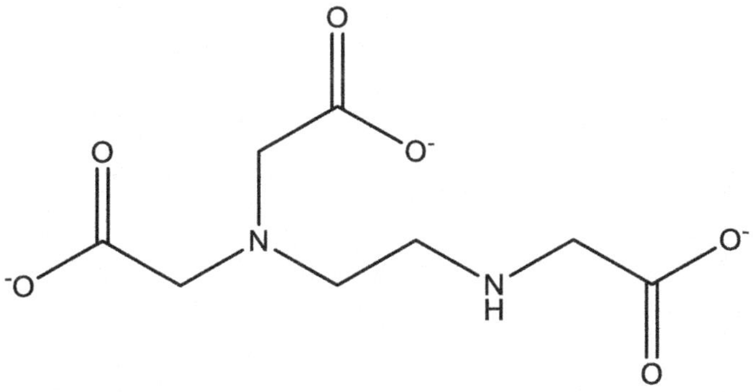

2,2'-((2-((carboxylatomethyl)amino)ethyl)azanediyl)diacetate

Hexadentate Ligands

If a ligand carries six donor atoms in its structure, it is called a hexadentate ligand. Such ligand possesses six coordination sites available.

Ethylenediaminetetraacetate ligand is a very famous hexadentate ligand.

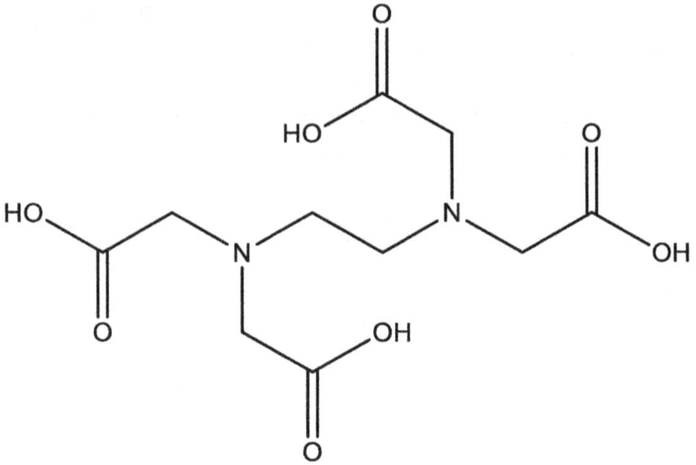

2,2',2'',2'''-(ethane-1,2-diylbis(azanetriyl))tetraacetic acid

NOMENCLATURE

You can name the Transition metal complexes trivially or according to a set of rules by IUPAC. Trivial names are not uniform and may differ from one place to other. IUPAC nomenclature provides a set of rules that makes the terminology systematic and consistent across the scientific community.

$[Fe^{II}(CN)_6]^{4-}$	$[Fe^{III}(CN)_6]^{3-}$
Ferrocyanide (Trivial Name)	Ferricyanide (Trivial Name)
Hexacyanoferrate(II) (IUPAC Name)	Hexacyanoferrate(III) (IUPAC Name)

In this book, we shall discuss how to write the names of transition metal complexes according to IPAC nomenclature. You will learn the best if you follow all the sections step by step.

Ruel 1:

Transition metal complexes may comprise two parts - a cationic part and an anionic part. Write the name of the cationic part first, and the anionic part follows.

$$K_4[Fe^{II}(CN)_6]$$

Potassium hexacyanoferrate(II)

In the above example, potassium being the cationic (positive) part is named first, whereas the hexacyanoferrate(II) being the anionic (negative) part is named later.

Rule 2:

Write the names of ligands before the names of the metal atoms.

$K_4[Fe^{II}(CN)_6]$

Potassium hexacyanoferrate(II)

In the above example, CN is the ligand. Therefore, we have written the name cyano before the metal name.

Rule 3:

To write the names of the ligands, we need to take care of the following sub-rules.

Rule 3a:

The names of anionic ligands (ligands carrying negative charges) may end in -o, -ato, -ido, -ito, etc.

CN^- = cyano

CH3COO$^-$ = acetate

ONO^- = nitrito

Cl^- = chloro or chlorido

Br^- = bromido

$NH_2CH_2COO^-$ = glycinato

I- = Iodo

F^- = fluoride

NH_2^- = amido

NH^{2-} = imido

N_3^- = nitrido

P_3^- = phosphide

HS^- = hydrogen sulphido, etc.

SO_3^{2-} = sulphito

Rule 3b:

For neutral ligands, there are no systematic names. NH_3 is named as ammine, H_2O as aqua, CH_3NH_2- as methylamine, PH_3 as phosphine, etc.

Rule 3c:

The names of organic free radicals are often written unchanged. For example, C_2H_5 is written as ethyl, C_6H_5 is written as phenyl, etc.

Rule 3d:

Monodentate ligands which contain more than two donor atoms in their structure are called ambident ligands. Such ligands can attach to the metal atom through either of the donor atoms. The attached donor atom is specified by

- Using a specific name. For example, in the case of CN, cyano when C is the attaching atom and isocyano if N is the attaching atom.
- The attaching atom is denoted by a bar over the atom symbol. For example, for CN, if the attaching atom is C, then place a bar over C.

Rule 3e:

The ligands are named alphabetically. For example, if there are two ligands, aqua, and chloro attached to the same metal atom, aqua will be named first because it starts with 'a' whereas chloro starts with 'c'. See the below example

$[Pt(H_2O)_4Cl_2]$ is named tetraaquadichloroplatinum(II)

Note that the prefixes di, tri, tetra are not considered while assigning the order of the ligands.

Rule 3f:

If more than one ligand of the same type is attached to the metal atom, prefixes di, tri, tetra, etc., are used to define the number of the same ligands.

For example,

$[Pt(H_2O)_4Cl_2]$ tetraaquadichloroplatinum(II).

In this example, there is four water (aqua molecules attached to the metal atom. Hence, the prefix 'tetra' is used to denote the number of aqua ligands. Similarly, a prefix 'di' indicates two chloro ligands attached to the metal atom.

Rule 3g

In case chelating ligands are attached with the metal atom that contains the 'di,' 'tri' already in their names, it would be confusing to reuse such prefixes again. For example, ethylenediamine already has 'di' in its name. Hence, using diethylenediamine would be confusing to name if two molecules are attached to the metal atom.

In such cases, the prefix bis, tris, tetrakis, etc., are used to denote the same chelating ligand number.

For example,

$[CoCl_2(en)_2]Cl_2$ is named as dichloro-bis-ethylenediamminecolbalt(IV) chloride.

Rule 3h

If a ligand attaches simultaneously with more than one metal atom, it is called a bridging ligand. Complexing containing bridging ligands are called bridge complexes.

A Greek letter -*u* is used for indicating the bridging ligand.

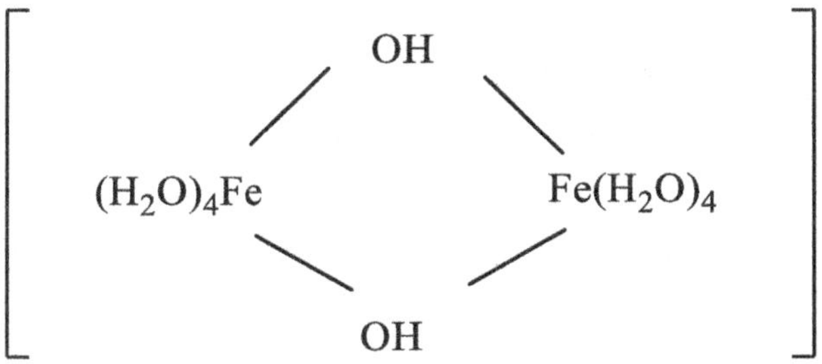

octaaqua-*u*-dihydroxo-di-iron(III) sulphate

Rule 4:

The name of a metal atom in the complex depends on the formal charge on the complex part (within the square brackets).

If the complex part carries a negative charge, a suffix -ate is added with the metal atom's name.

For example, ferrate, cobaltate, etc.

$K_2[Fe(CN)_4]$ Potassium hexacyanoferrate(II)

Since the complex part, $[Fe(CN)_4]$ carries a negative charge. On the other hand, No such addition is required if the complex part has a positive charge or is a neutral part.

$[Co(NH_3)_4Cl_2]SO_4$ tetraamminedichlorocobalt(IV) sulphate

Here, the complex part, $[Co(NH_3)_4Cl_2]$, carries a positive charge.

Now that you have learned all the necessary rules to name transition metal complexes, let's see examples to practice.

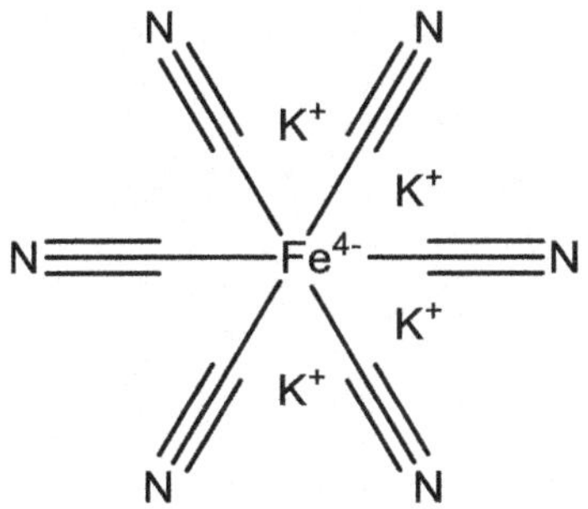

$K_4[Fe(CN)_6]$

potassium hexacyanoferrate(II)

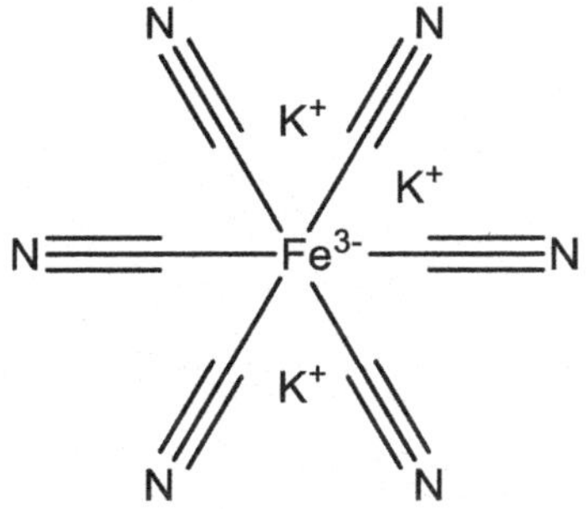

$K_3[Fe(CN)_6]$

potassium hexacyanoferrate(III)

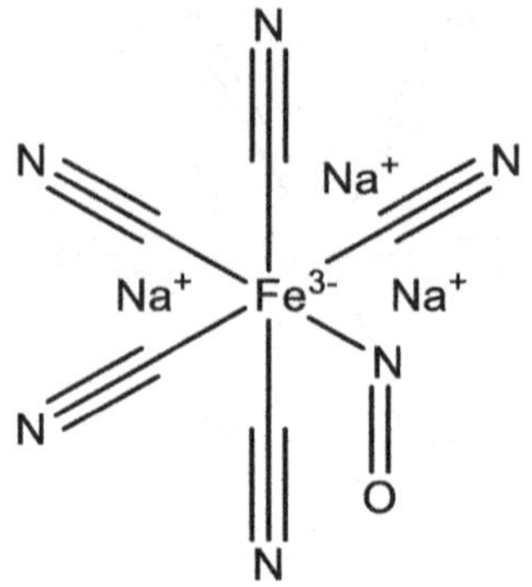

$Na_3[Fe(CN)_5NO]$

sodium pentacyano(nitroso)ferrate(III)

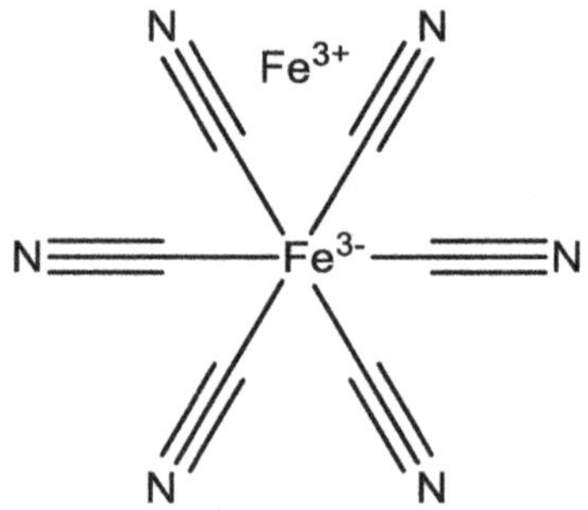

$Fe[Fe(CN)_6]$

iron(III) hexacyanoferrate(III)

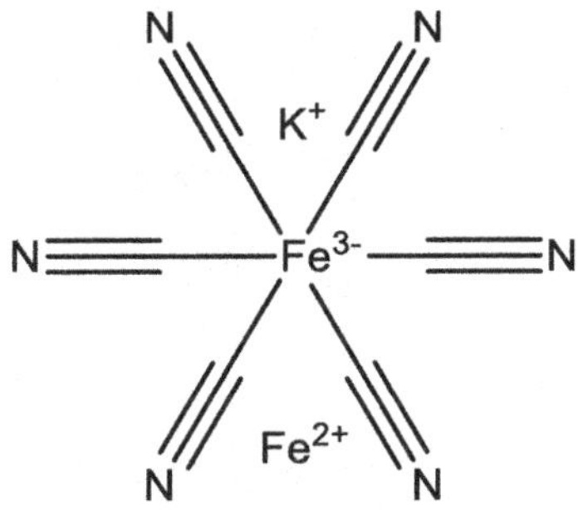

$KFe[Fe(CN)_6]$

iron(II) potassium hexacyanoferrate(III)

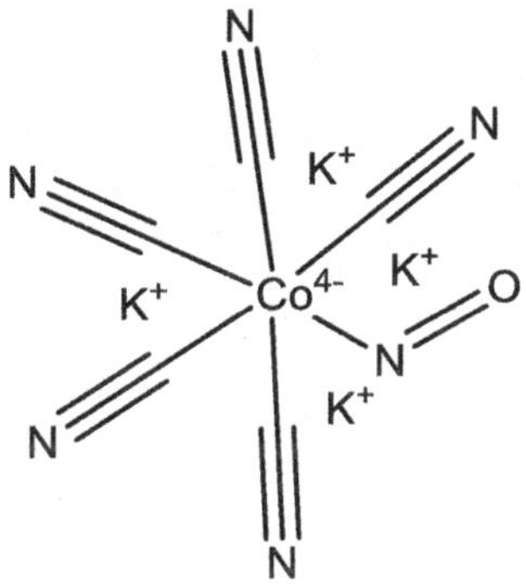

$K_4[Co(CN)_5NO)]$

potassium pentacyano(nitroso)cobaltate(II)

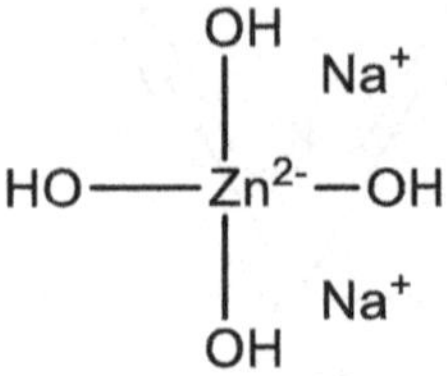

$Na_2[Zn(OH)_4]$

sodium tetrahydroxyzincate(II)

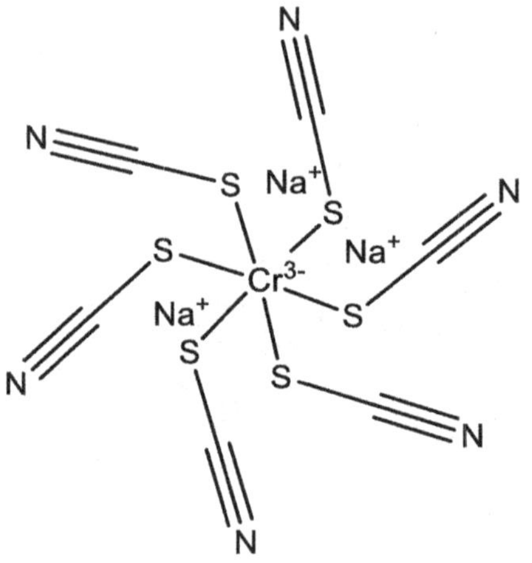

$Na_3[Cr(SCN)_6]$

sodium hexathiocyanatochromate(III)

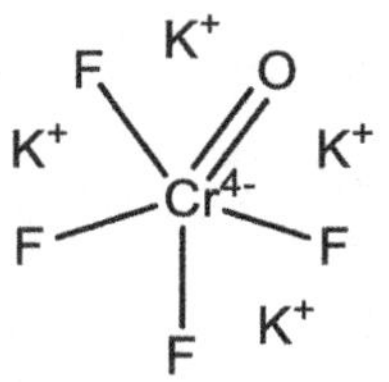

$K_4[CrF_4O]$

potassium tetrafluoro(oxo)chromate(II)

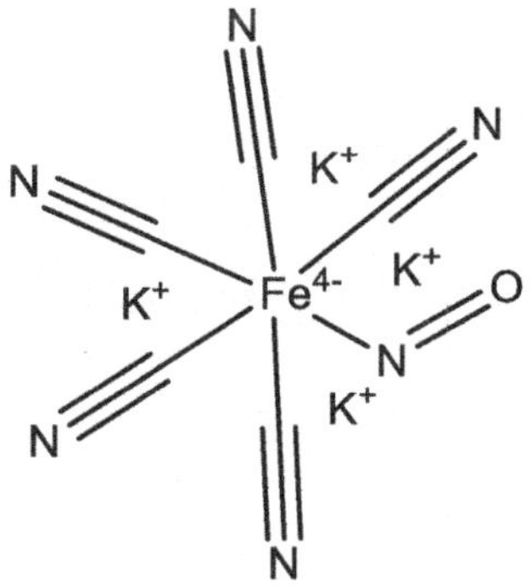

$K_4[Fe(CN)_5(NO)]$

potassium pentacyano(nitroso)ferrate(II)

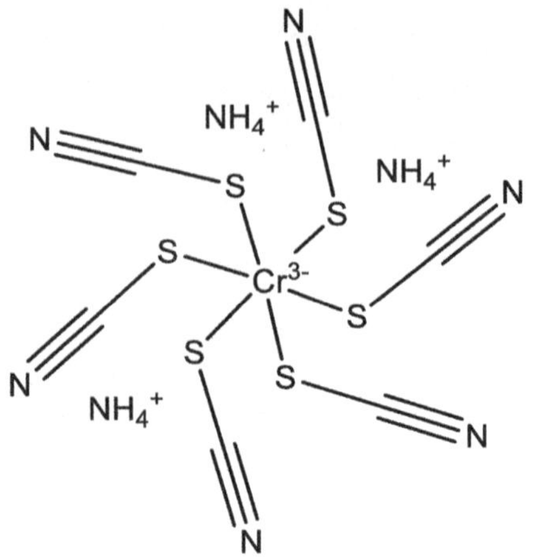

$(NH_4)_3[Cr(SCN)_6]$

ammonium hexathiocyanatochromate(III)

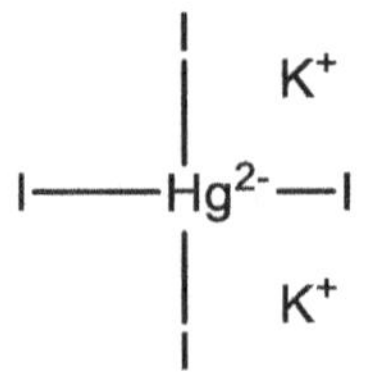

$K_2[HgI_4]$

potassium tetraiodomercurate(II)

ISOMERISM IN COORDINATION COMPOUNDS

If two or more compounds have the same chemical formula but different structures, such compounds are called isomers, and this phenomenon is known as isomerism.

Coordination compounds also show isomerism. Here, we shall learn various kinds of isomerism present in coordination complexes.

Structural Isomerism

Structural isomers differ in how the atoms are bonded together. For example,

H-N-C=O N=C-O-H

Both ligands have a similar chemical formula, but atoms' linkage is different in the two ligands. Hence, we shall say that these isomers are structural isomers.

Ionization Isomerism

These types of isomers differ in the ionizable groups. These isomers may have interchanged ions outside and inside the coordination sphere.

For example,

$[Pt(H_2O)_4Cl_2]Br_2$ $[Pt(H_2O)_4Br_2]Cl_2$

Both these complexes differ in the ionizable ions.

Hence, there are called ionization isomers.

Hydrate Isomerism

Hydrate isomers are different from each other in terms of water molecules present inside and outside the coordination sphere. In simple words, these isomers are different from each other in the placement of water molecules.

Examples are

$[Cr(H_2O)_6]Cl_3$	violet color
$[Cr(H_2O)_5Cl]Cl_2.H_2O$	light green color
[Cr(H2O)4Cl2]Cl.2H2O	dark green

Linkage Isomers

These isomers arise because a ligand can coordinate with the metal atom with more than one atom present in its structure.

For example,

$[Co(NH_3)_5(ONO)]Cl_2$

In this complex, the ligand -ONO is attached with the metal atom through its -O atom.

$[Co(NH_3)_5(NO_2)]Cl_2$

In this complex, the ligand NO2 is attached to the metal atom.

Coordination Sphere Isomerism

There are many examples of coordination compounds where both the cationic and anionic parts of the complex are complex. Therefore, coordination sphere isomerism may arise due to the exchange of ligands between the cationic complex and the anionic complex.

$[Pt(H_2O)_4][PtCl_6]$

and

$[Pt(H_2O)_4Cl_2][PtCl_4]$

Notice here that the ligands are exchanged between the cationic part and anionic part of the complex. This results in different isomers.

Another example is

$[Cu(H_2O)_4][PtCl_4]$

and

$[Pt(H_2O)_4][CuCl_4]$

Coordination Position Isomerism

In the case of polynuclear complexes, it becomes possible that ligands may exchange within the coordination sphere.

See the below example to understand the concept

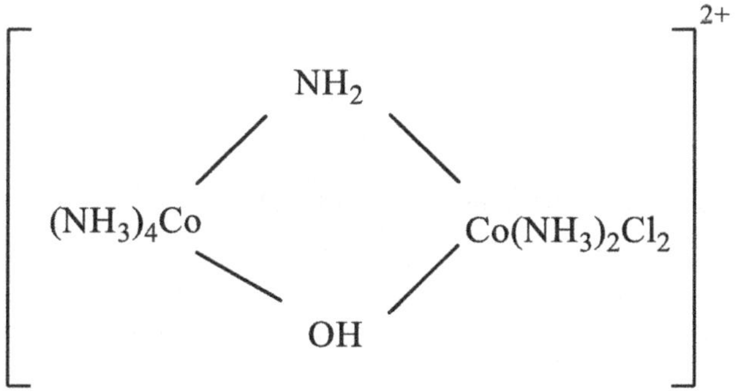

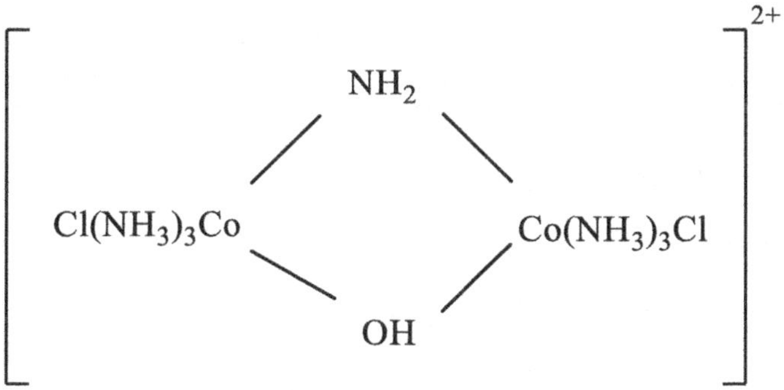

Ligand Isomerism

Some ligands can show isomerism themselves. If such ligands are present in the complex, we can see the isomerism due to these ligands. For example,

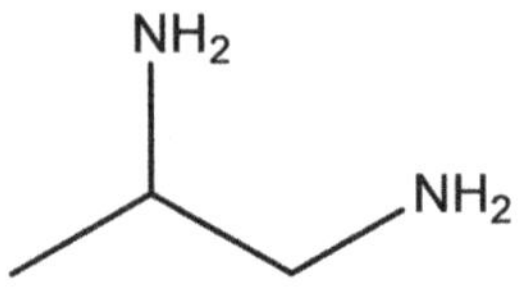

propane-1,2-diamine

propane-1,3-diamine

1,2-diaminopropane and 1,3-diaminopropane are isomers. If these ligands are present in two complexes, the complexes will be ligand isomers.

Conformation Isomerism

Such isomers are identical to each other, except there is a difference in geometries of the ligands.

For example,

$[Ni(P.Et.Ph_2)_2Br_2$	tetrahedral	green in color
$[Ni(P.Et.Ph_2)_2Br_2$	square planer	brown in color

The tetrahedral isomer is paramagnetic in nature, whereas the square planer isomer is diamagnetic.

Stereoisomers

These are isomers with the same formula, but the ligands are arranged differently in space.

Such isomers are called stereoisomers.

Geometrical Isomerism

In geometrical isomerism, ligands occupy different positions around the central metal atom.

If similar ligands are present on one side of the metal atom, such isomer is a -cis isomer.

If similar ligands are present on the opposite side of the metal atom, such isomer is called a -trans isomer.

Geometrical isomerism depends on the coordination number of the complex. Let's see different geometrical isomers with different coordination numbers.

(Coordination number is the total number of ligands attached with the central metal atom).

Geometric isomerism in complexes with coordination number 4

Coordination complexes with coordination number four exist in two geometric types.

(i) Tetrahedral geometry
(ii) Square planer geometry

Geometrical isomerism in tetrahedral complexes

Tetrahedral complexes with coordination number four do not show geometric isomerism.

Geometrical isomerism in square planer complexes

The geometries of square planer complexes can be classified as

$[Ma_4]$ type
- $[Ma_3b]$ type
- $[Mab_3]$ type
- $[Ma_2b_2]$ type
- $[Mabcd]$ type
- $[Ma_2bc]$ type

Here a, b, c, and d represent different ligands attached to the central metal atom.

The first three $[Ma_3b]$ types, $[Mab_3]$ type and $[Mab_3]$ type, do not show geometrical isomerism because all variants become similar.

The other types show geometrical isomerism that is explained further.

[Ma_2b_2] type

In this complex, cis and trans isomerism exists. Here M represents the central metal atom. 'a' and 'b' represent ligands.

For example, [$Pt(NH_3)_2Br_2$] can exist as *cis*-[$Pt(NH_3)_2Br_2$] and *trans*-[$Pt(NH_3)_2Br_2$] isomers.

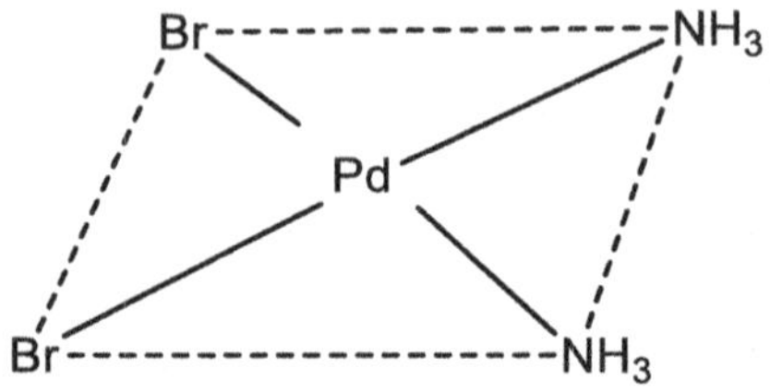

cis-isomer

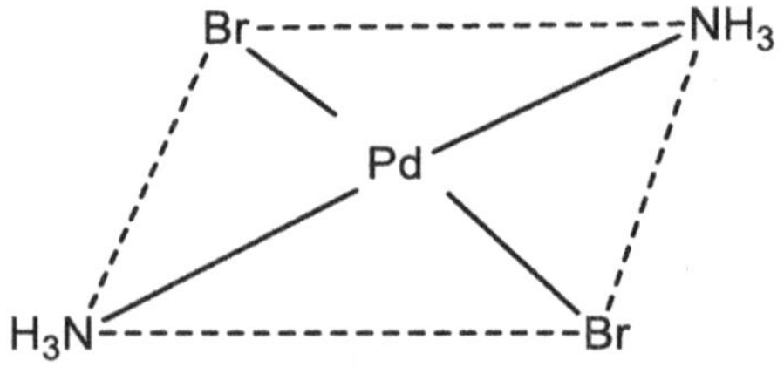

trans-isomer

[Ma2bc] type

Cis- and *trans*- isomerism is also possible in [Ma_2bc] type coordination complexes.

The complex [$PdClBr_2I$] exists as *cis*- and *trans*- isomers

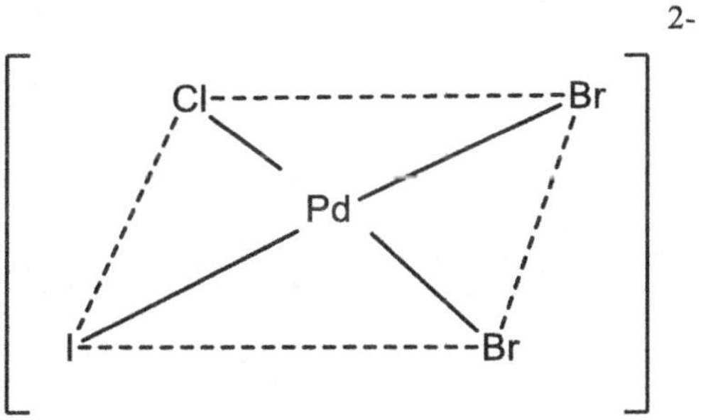

cis-isomer

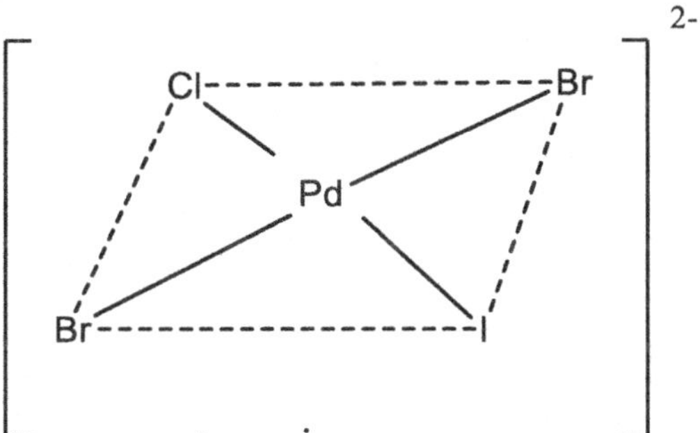

trans-isomer

Similarly

The complex [$Pt(NH_3)_2BrNO_2$] exists as *cis*- and *trans*- isomers

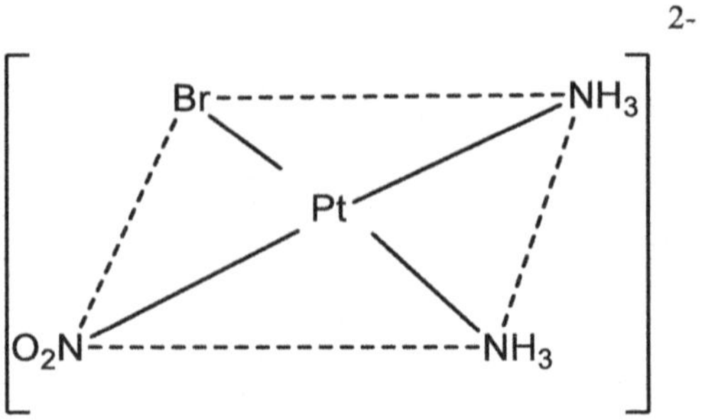

cis-isomer

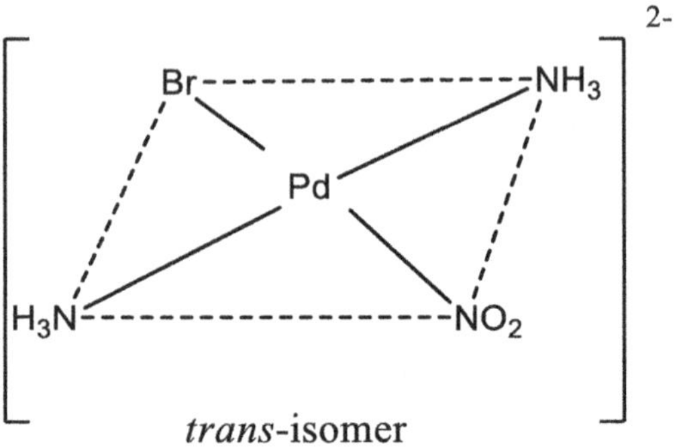

trans-isomer

[Mabcd] type

Complexes of this type have all four different ligands. The following example expresses isomerism in such complexes.

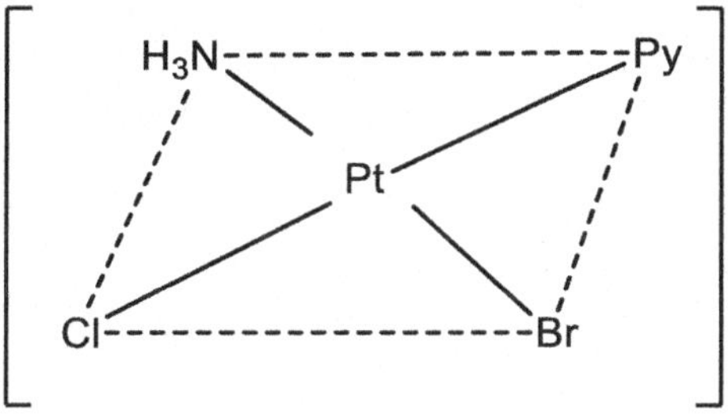

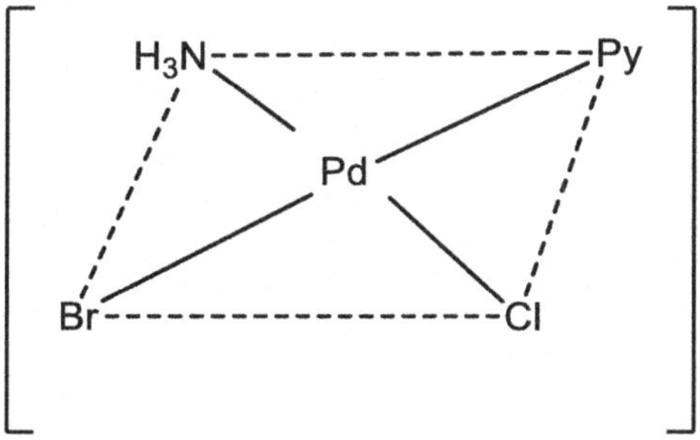

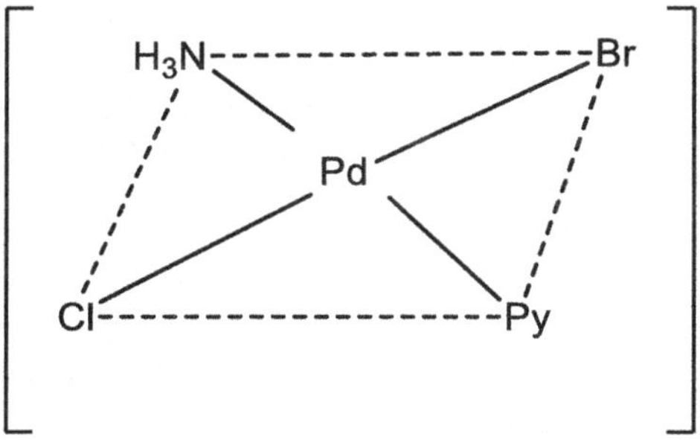

This complex can exist in three isomers.

Geometrical isomerism in bridged complexes

Bridged complexes can exist in different isomers. For example,

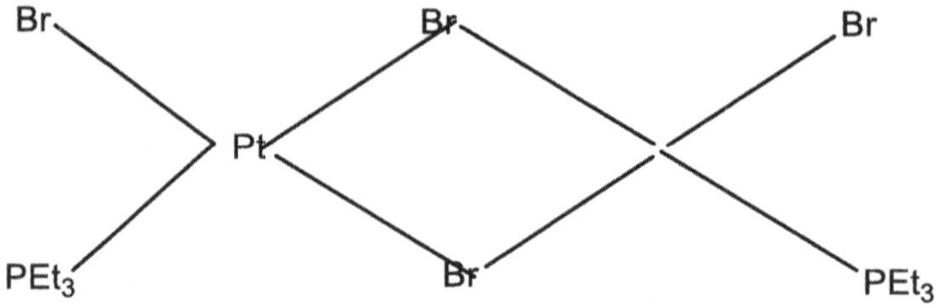

cis-isomer

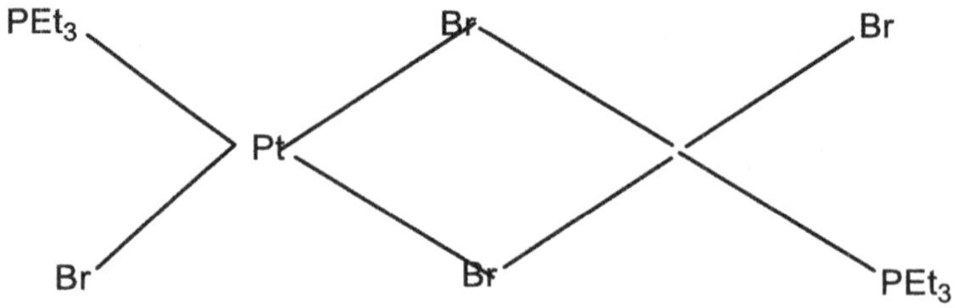

trans-isomer

Geometrical isomerism in octahedral complexes

Octahedral complexes also demonstrate geometrical isomerism. Examples are given below

[Ma_4b_2] type octahedral complexes

[Ma_4b_2] type octahedral complexes show cis and trans isomerism.

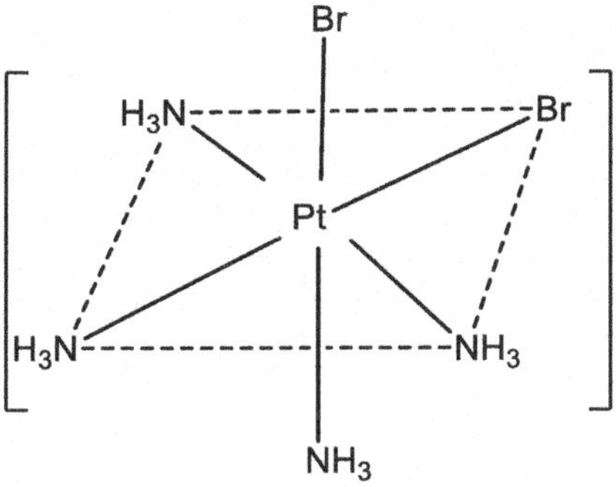

cis-isomer

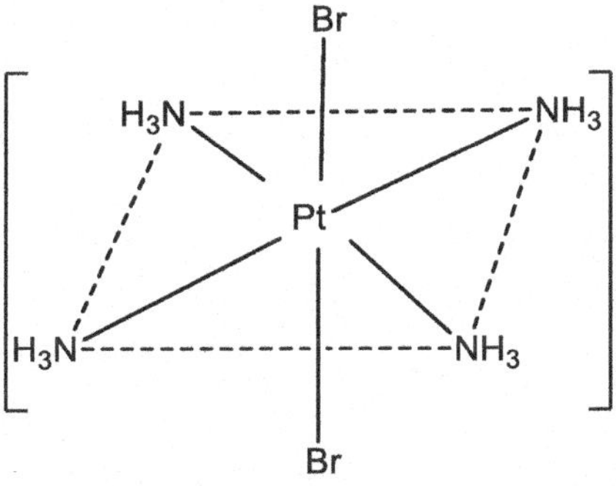

trans-isomer

[Ma_3b_3] type octahedral complexes

[Ma_3b_3] type octahedral complexes show cis and trans isomerism.

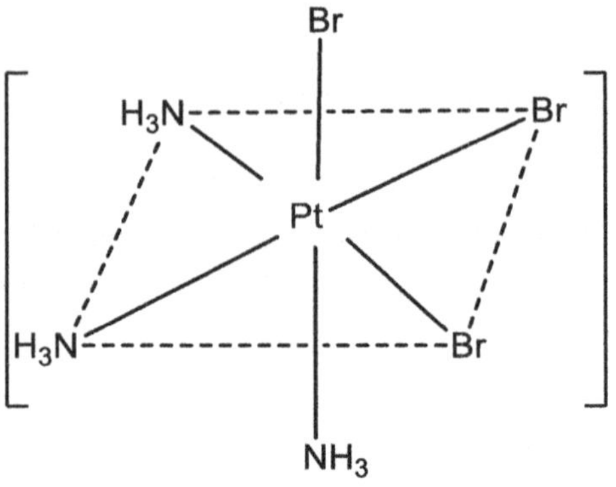

cis-isomer

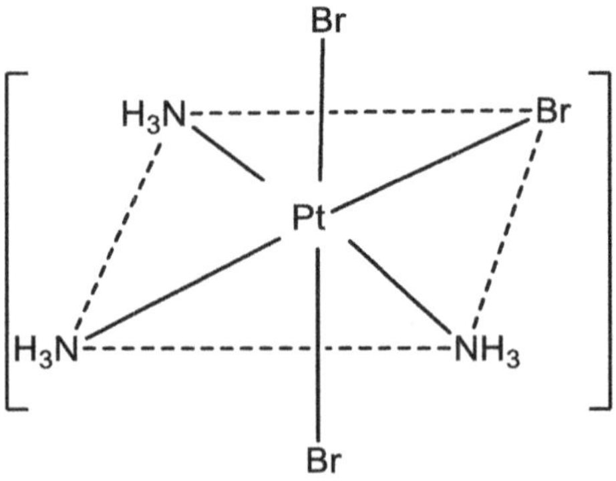

trans-isomer

$[Ma_2b_2C_2]$ type octahedral complexes

$[Ma_3b_3]$ type octahedral complexes can also exist as isomers

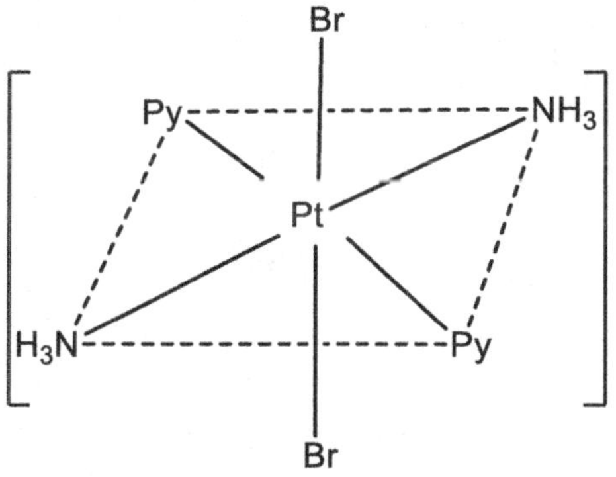

(I)

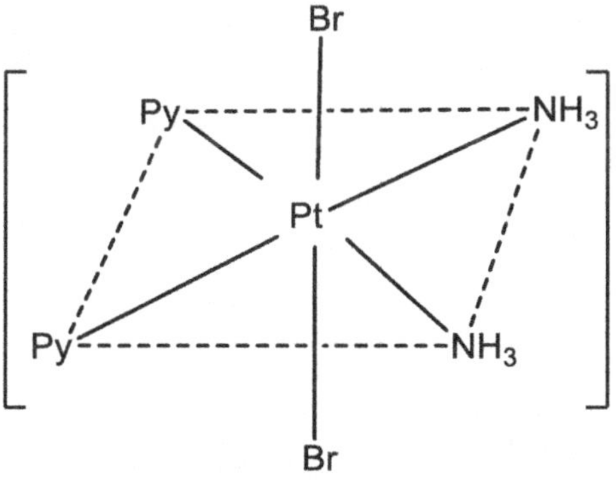

(II)

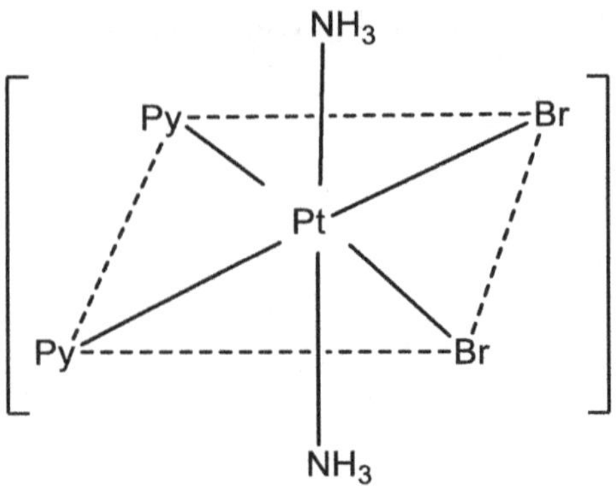

(III)

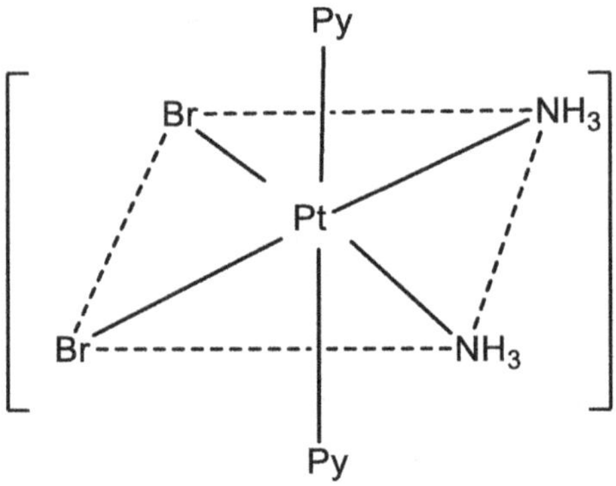

(IV)

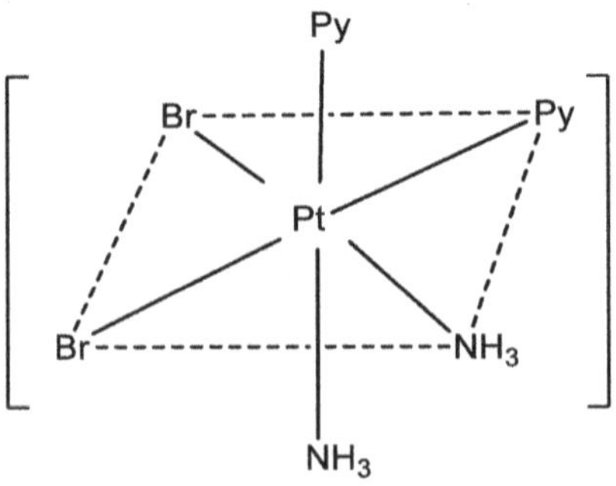

(V)

[M(AA)$_2$ab] type octahedral complexes

These complexes also show *cis*- and *trans*- isomerism. Here (AA)2 represents a bridging ligand.

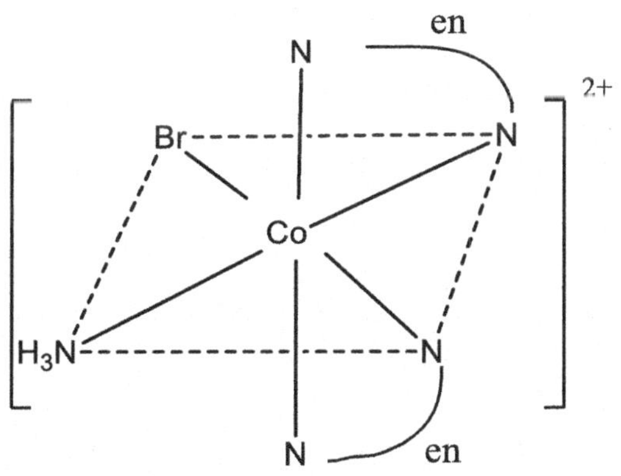

cis-isomer

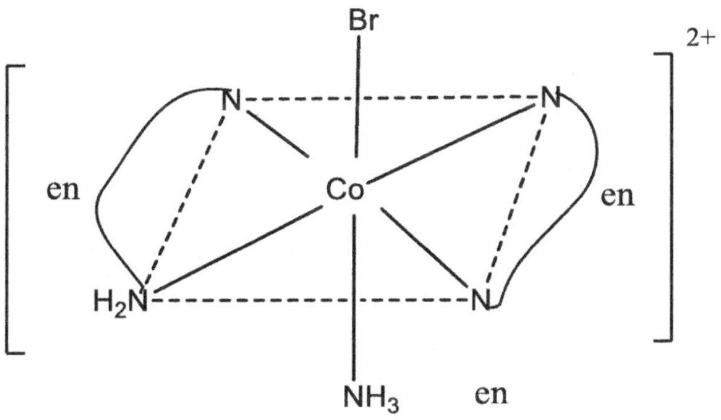

trans-isomer

[M(AB)3] type

Octahedral complexes of [M(AB)3] type can show *cis-* and *trans-* isomerism. For this type of geometrical isomerism, there is a condition that the bidentate ligand has two different donor atoms.

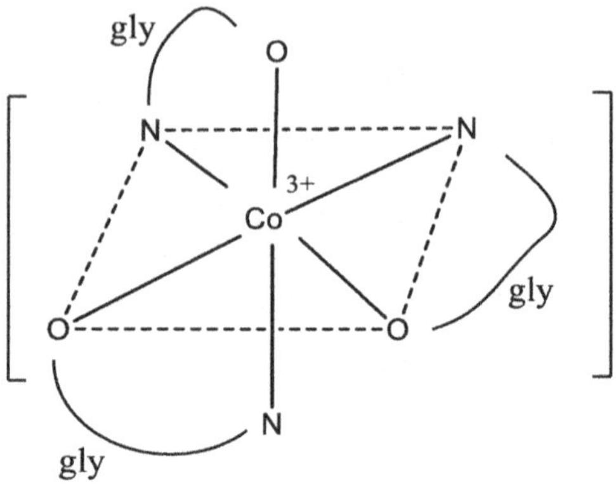

cis-isomer

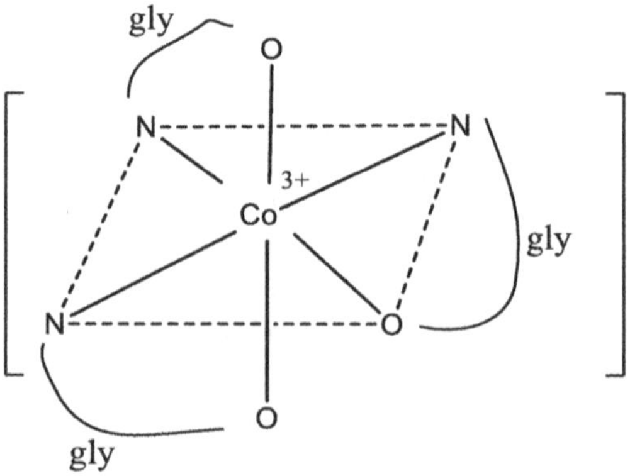

trans-isomer

OPTICAL ISOMERISM

Existing of isomers as non-superimposable mirror images is called optical isomerism.

Octahedral complexes can exist in dextro-rotatory and levo-rotatory isomers.

Few types of octahedral complexes show optical isomerism.

[$Ma_2b_2c_2$] type octahedral complexes

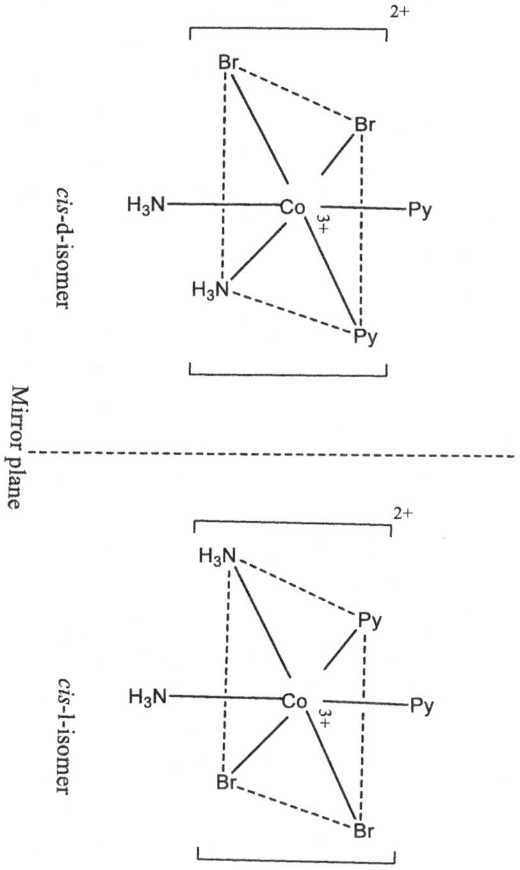

[$Ma_2b_2c_2$] type octahedral complexes

Octahedral complexes of this type also show optical isomerism.

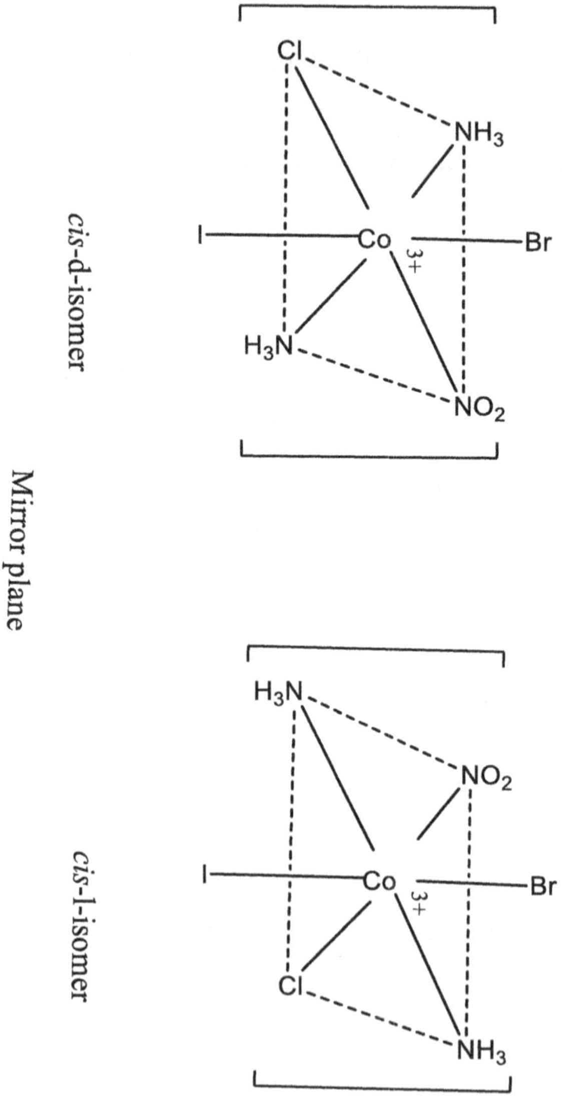

The above-given examples are to introduce the concept. More details can be read from the suggested books.

References for further reading

1. Inorganic Chemistry
Willer, Overton, Rourke, Armstrong

2. Advanced Inorganic Chemistry
F. Albert Cotton, Geoffrey Wilkinson, Carlos A. Murillo

ABOUT THE AUTHOR

Dr. Muhammad Jawwad Saif is an associate professor at the Department of Applied Chemistry, Government College University Faisalabad – Pakistan.

Made in the USA
Monee, IL
08 November 2026

67967617R00038